What's the Code?

AF676029

Example

Use the code to find the product.

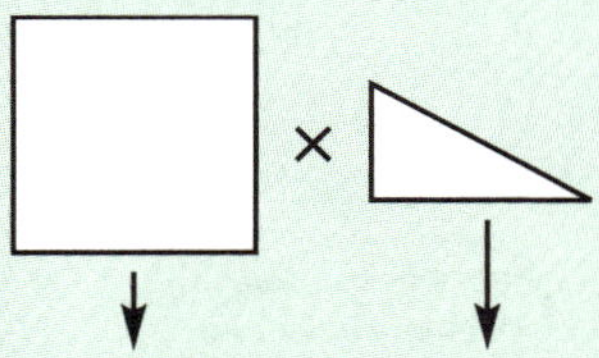

9 × 4 = **36**

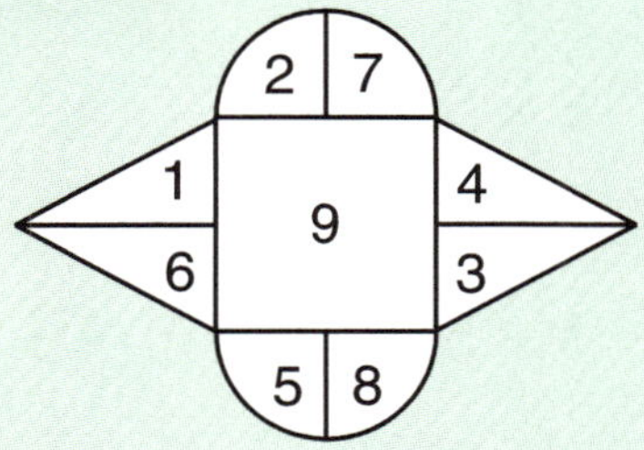

Use the code to find the product.

1. 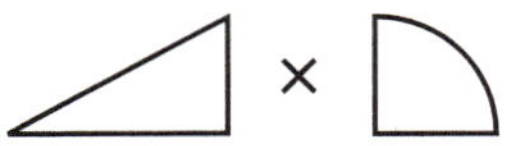

2. 

3. 

4. 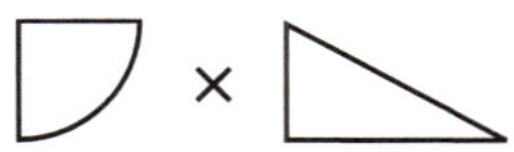

5. 

6. 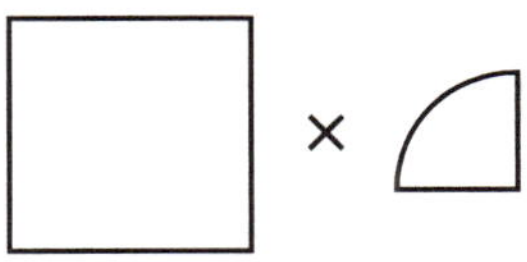

7. 

8. 

9. 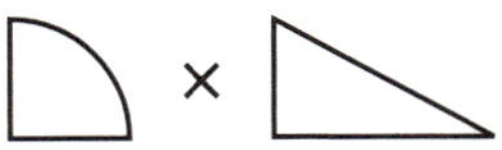

10. 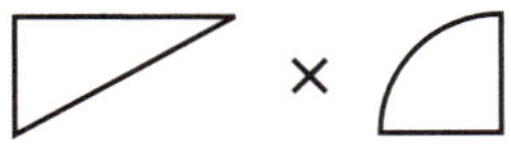

11. 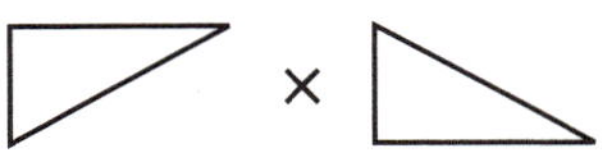

12. 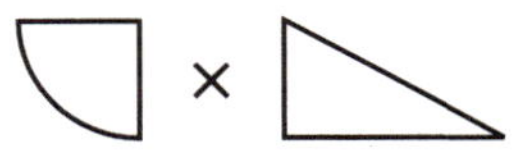

Answer Box

A	B	C	D	E	F
28	16	14	20	7	10
G	**H**	**I**	**J**	**K**	**L**
12	24	3	18	6	32

Objective: Use a code to find a product with a factor of 1, 2, or 4.

The Fives Factor and More

Find the product.

1. 6 × 4
2. 1 × 6
3. 5 × 5
4. 9 × 6
5. 8 × 5
6. 6 × 8
7. 3 × 5
8. 6 × 6
9. 5 × 6
10. 6 × 3
11. 4 × 5
12. 5 × 9

Answer Box

A	B	C	D	E	F
24	54	6	48	30	25
G	**H**	**I**	**J**	**K**	**L**
15	18	45	36	40	20

Find the product.

1. 5 × 4
2. 9 × 6
3. 5 × 2
4. 6 × 8
5. 4 × 6
6. 5 × 7
7. 6 × 6
8. 6 × 5
9. 2 × 6
10. 6 × 7
11. 5 × 3
12. 5 × 9

Answer Box

A	B	C	D	E	F
45	30	12	24	36	42
G	**H**	**I**	**J**	**K**	**L**
54	35	15	10	20	48

Objective: Find a product with a factor of 5 or 6.

Three and Nine to the Rescue!!

Find the product.

1. 4 × 3
2. 9 × 9
3. 6 × 3
4. 9 × 1
5. 3 × 8
6. 7 × 9
7. 8 × 9
8. 3 × 2
9. 3 × 7
10. 5 × 3
11. 6 × 9
12. 4 × 9

Answer Box

A	B	C	D	E	F
63	81	9	12	18	21
G	**H**	**I**	**J**	**K**	**L**
6	36	15	72	54	24

Find the total cost.

1. 2 × 3¢
2. 5 × 9¢
3. 3 × 3¢
4. 8 × 9¢
5. 7 × 9¢
6. 4 × 3¢
7. 2 × 9¢
8. 9 × 3¢
9. 8 × 3¢
10. 7 × 3¢
11. 5 × 3¢
12. 4 × 9¢

Answer Box

A	B	C	D	E	F
24¢	63¢	27¢	36¢	21¢	72¢
G	**H**	**I**	**J**	**K**	**L**
15¢	9¢	45¢	12¢	6¢	18¢

Objective: Find a product with a factor of 3 or 9.

Just the Facts, Please

Find the product.

1. 7 × 4	2. 9 × 7	3. 1 × 7
4. 8 × 1	5. 3 × 8	6. 7 × 7
7. 8 × 8	8. 8 × 2	9. 5 × 7
10. 7 × 8	11. 6 × 7	12. 8 × 9

Answer Box

A	B	C	D	E	F
35	7	63	49	28	8
G	**H**	**I**	**J**	**K**	**L**
42	24	16	72	56	64

Find the factors for the product.

1. 14	2. 42	3. 56	4. 32
5. 21	6. 48	7. 24	8. 63
9. 49	10. 40	11. 72	12. 16

Answer Box

A	B	C	D	E	F
7 × 2	8 × 4	7 × 3	8 × 9	7 × 6	6 × 8
G	**H**	**I**	**J**	**K**	**L**
8 × 3	8 × 5	8 × 7	7 × 7	8 × 2	7 × 9

Objective: Find a product with a factor of 7 or 8.

Something's Missing

Find the missing factor or product.

1. $6 \times 8 = \blacksquare$
2. $4 \times \blacksquare = 36$
3. $6 \times \blacksquare = 30$
4. $3 \times 8 = \blacksquare$
5. $9 \times \blacksquare = 72$
6. $\blacksquare \times 8 = 16$
7. $\blacksquare \times 3 = 18$
8. $\blacksquare \times 5 = 35$
9. $\blacksquare \times 5 = 15$
10. $8 \times 5 = \blacksquare$
11. $7 \times 3 = \blacksquare$
12. $6 \times \blacksquare = 24$

Answer Box

A	B	C	D	E	F
40	6	7	4	21	8
G	**H**	**I**	**J**	**K**	**L**
24	48	2	9	5	3

Objective: Find a missing factor or product to replace a variable.

Let's Do It in Threes

Example

Use the code to find the product.

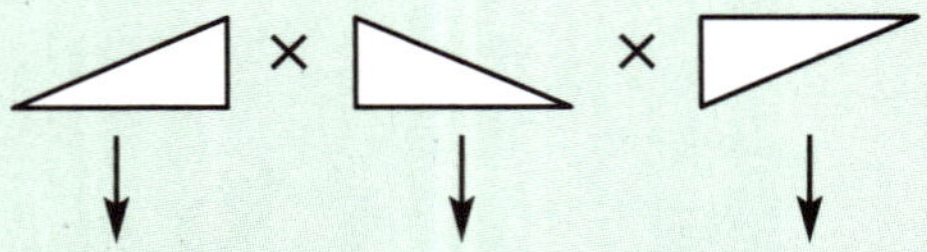

1 × 2 × 5 = **10**

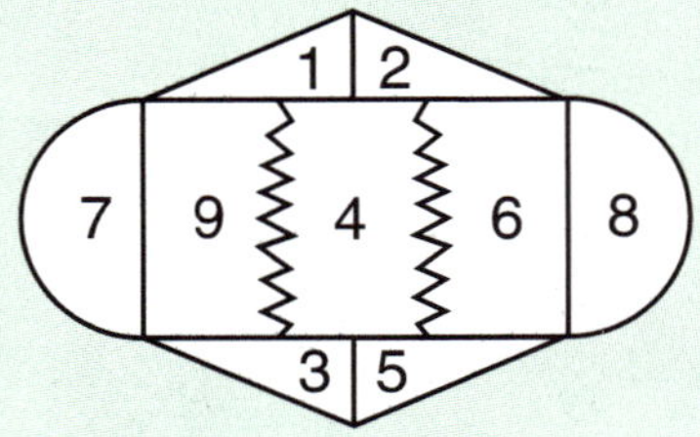

Use the code to find the product.

1. 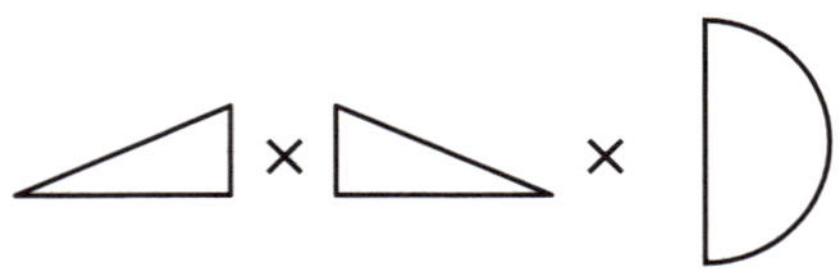
2. 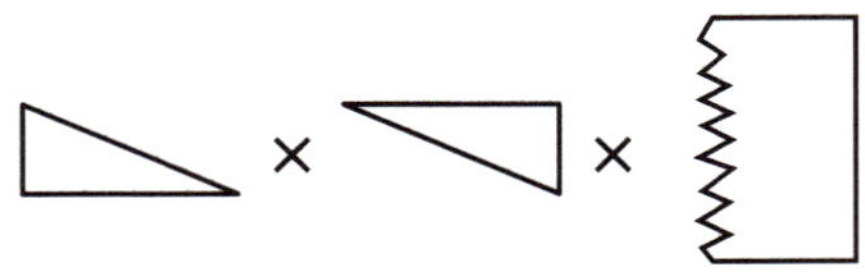
3. 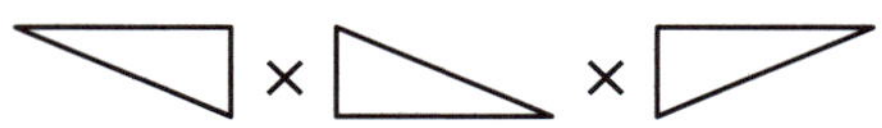
4. 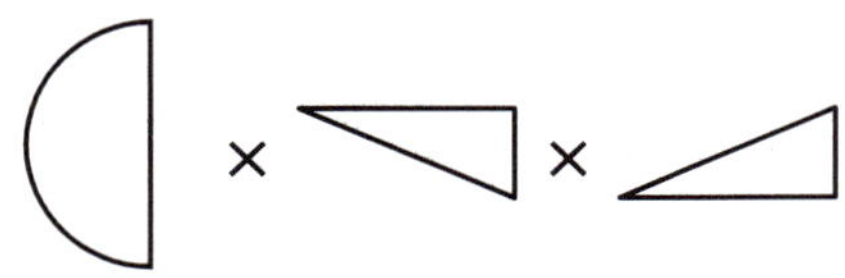
5. 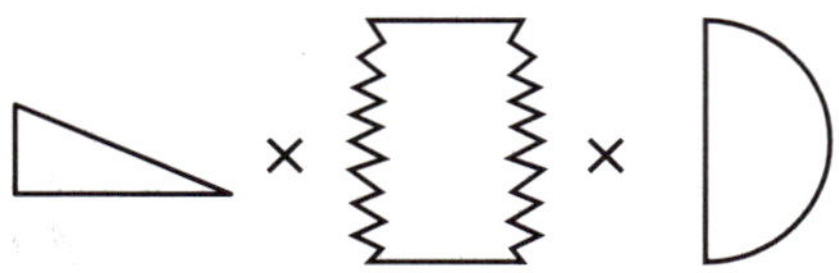
6. 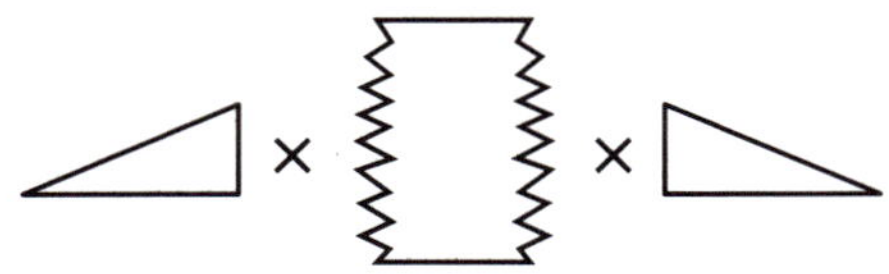
7. 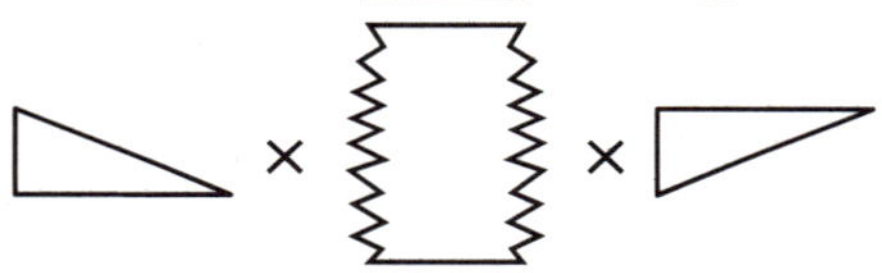
8. 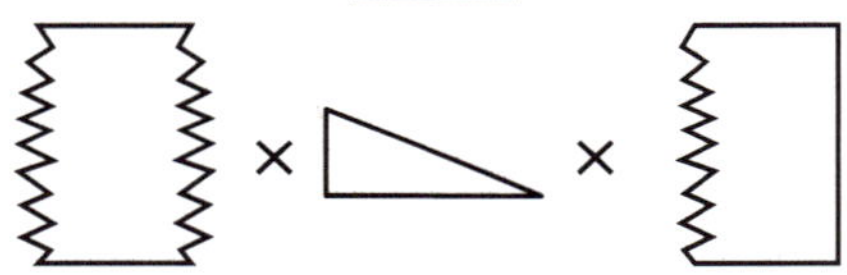
9. 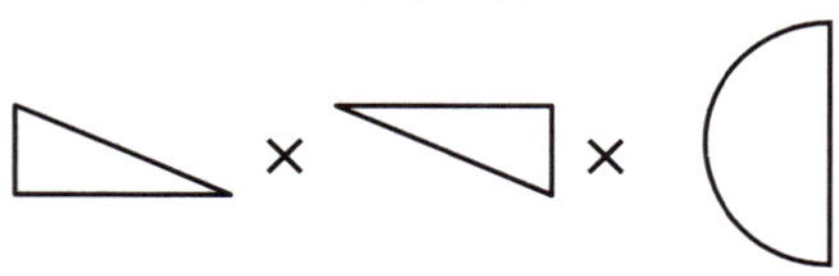
10. 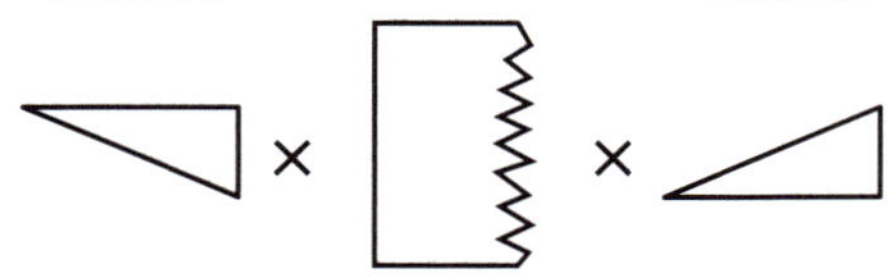
11. 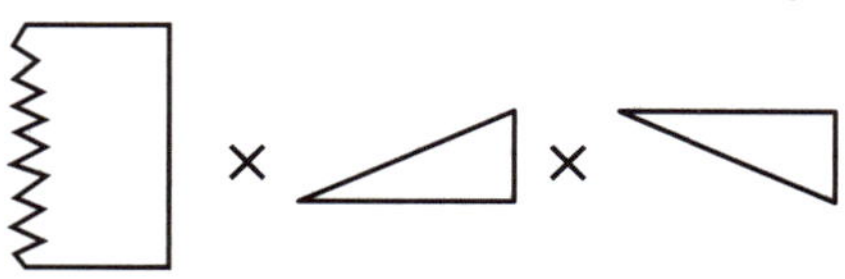
12. 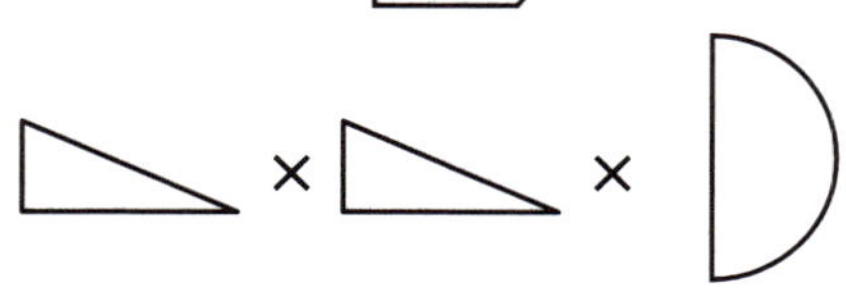

Answer Box

A	B	C	D	E	F
8	48	21	27	30	18
G	**H**	**I**	**J**	**K**	**L**
32	36	40	16	64	42

 Objective: Use a code to find the product of three 1-digit numbers.

What's First on the List?

Find the answer.

1. (3 × 6) + 4
2. (6 × 8) + 6
3. (4 × 7) + 2
4. (10 × 7) + 7
5. (6 × 2) + 7
6. (6 × 9) + 2
7. (7 × 4) + 7
8. (9 × 9) + 15
9. (8 × 3) + 4
10. (3 × 9) + 5
11. (7 × 5) + 8
12. (5 × 9) + 6

Answer Box

A	B	C	D	E	F
43	30	32	77	96	56
G	**H**	**I**	**J**	**K**	**L**
28	19	22	35	54	51

Objective: Find an answer by multiplying and then adding.

No Change Here!

The Order Property of Multiplication

Changing the order of the factors does not change the product.

$9 \times 8 = 8 \times 9$

Find the missing factor.

1. $4 \times 6 = 6 \times \blacksquare$
2. $8 \times 11 = \blacksquare \times 8$
3. $4 \times \blacksquare = 10 \times 4$
4. $3 \times \blacksquare = 2 \times 3$
5. $7 \times 5 = 5 \times \blacksquare$
6. $7 \times 6 = \blacksquare \times 7$
7. $8 \times 3 = \blacksquare \times 8$
8. $\blacksquare \times 5 = 5 \times 1$
9. $7 \times 12 = \blacksquare \times 7$
10. $\blacksquare \times 4 = 4 \times 9$
11. $\blacksquare \times 9 = 9 \times 5$
12. $8 \times 7 = 7 \times \blacksquare$

Answer Box

A	B	C	D	E	F
3	9	7	5	8	1
G	**H**	**I**	**J**	**K**	**L**
4	2	12	10	11	6

Objective: Use the order property to find a missing factor.

Property Rights

The Grouping Property of Multiplication

The way in which you group factors does not change the product.

$$2 \times (3 \times 4) = (2 \times 4) \times 3$$

Find the missing factor.

1. 4 × 5 × 7 = (4 × ■) × 5
2. 1 × 4 × 6 = 1 × (6 × ■)
3. 6 × 10 × 3 = (6 × 3) × ■
4. 4 × 3 × 2 = 4 × (2 × ■)
5. 7 × 2 × 9 = 7 × (9 × ■)
6. 3 × 12 × 5 = 5 × (3 × ■)
7. 5 × 3 × 1 = 5 × (■ × 3)
8. 5 × 3 × 6 = ■ × (3 × 5)
9. 9 × 4 × 11 = 9 × (■ × 4)
10. 8 × 3 × 9 = 3 × (■ × 9)
11. 4 × 6 × 5 = (4 × ■) × 6
12. 7 × 2 × 9 = 7 × (■ × 2)

Answer Box

A	B	C	D	E	F
6	9	5	2	8	1
G	**H**	**I**	**J**	**K**	**L**
4	12	11	10	7	3

Objective: Use the order and grouping properties to find a missing factor.

Factor-Finding Mission

Factors of 16

16 = **1** × **16** 16 = **2** × **8** 16 = **4** × **4**

The factors of 16 are 1, 2, 4, 8, and 16.

Find the missing factor.

1. 6 1, 2, ■, 6
2. 20 1, 2, 4, ■, 10, 20
3. 4 ■, 2, 4
4. 18 1, ■, 3, 6, 9, 18
5. 10 1, 2, 5, ■
6. 24 1, 2, 3, 4, ■, 8, 12, 24
7. 14 1, 2, 7, ■
8. 25 1, 5, ■
9. 12 1, 2, 3, ■, 6, 12
10. 27 1, 3, ■, 27
11. 15 1, 3, 5, ■
12. 28 1, 2, 4, ■, 14, 28

Answer Box

A	B	C	D	E	F
1	4	2	3	6	5
G	**H**	**I**	**J**	**K**	**L**
10	15	14	9	7	25

 Objective: Find the missing factor in a list of all the factors of a number.

That's Prime!

Choose the prime number.

1. 34, 23, 14
2. 25, 50, 71
3. 12, 16, 29
4. 30, 44, 73
5. 31, 28, 22
6. 51, 55, 59

Choose the number that is *not* prime.

7. 71, 16, 29
8. 28, 59, 53
9. 11, 23, 34
10. 55, 31, 7
11. 37, 22, 31
12. 13, 51, 73

Answer Box

A	B	C	D	E	F
16	23	31	34	51	71
G	**H**	**I**	**J**	**K**	**L**
73	55	29	22	59	28

Objective: Identify a prime number.

Patterns Are Here, Too

Example

Find the missing number.

5 × 3 = 15, so 5 × 30 = ■

You can use what you know about multiplication facts to find a product with tens.

So, 5 × 30 = **150**.

Find the missing number.

1. 4 × 6 = 24, so 4 × 60 = ■
2. 5 × 9 = 45, so 5 × 90 = ■
3. 5 × 2 = 10, so 5 × 20 = ■
4. 7 × 3 = 21, so 7 × 30 = ■
5. 5 × 7 = 35, so 50 × 7 = ■
6. 8 × 6 = 48, so 80 × 6 = ■
7. 8 × ■ = 56, so 8 × 70 = 560
8. 5 × ■ = 30, so 5 × 60 = 300
9. 4 × 8 = 32, so 4 × ■ = 320
10. 7 × 9 = 63, so 7 × ■ = 630
11. 8 × 7 = 56, so ■ × 70 = 560
12. 9 × 3 = 27, so ■ × 30 = 270

Answer Box

A	B	C	D	E	F
9	6	350	8	7	90
G	**H**	**I**	**J**	**K**	**L**
480	450	100	80	210	240

Objective: Find the product of tens multiplied by a 1-digit number.

More Patterns

Example

Find the missing number.

4 × 3 = 12, so 4 × 300 = ■

You can use what you know about multiplication facts to find a product with hundreds.

So, 4 × 300 = **1,200**.

4 × 3 = 12

4 × 30 = 120

4 × 300 = 1,200

Find the missing number.

1. 4 × 8 = 32, so 4 × 800 = ■
2. 6 × 9 = 54, so 6 × 900 = ■
3. 6 × 2 = 12, so 6 × 200 = ■
4. 9 × 3 = 27, so 9 × 300 = ■
5. 5 × 5 = 25, so 500 × 5 = ■
6. 8 × 8 = 64, so 800 × 8 = ■
7. 8 × ■ = 40, so 8 × 500 = 4,000
8. 5 × ■ = 30, so 5 × 600 = 3,000
9. 2 × 8 = 16, so 2 × ■ = 1,600
10. 2 × 9 = 18, so 2 × ■ = 1,800
11. 8 × 7 = 56, so ■ × 700 = 5,600
12. 9 × 9 = 81, so ■ × 900 = 8,100

Answer Box

A	B	C	D	E	F
1,200	8	2,700	3,200	6,400	6
G	**H**	**I**	**J**	**K**	**L**
2,500	800	5	900	9	5,400

Objective: Find the product of hundreds multiplied by a 1-digit number.

About How Much Is That?

Example

Find the best estimate.

5 × 34 and 2 × 289

You can round one factor to make it easier to multiply mentally.

5 × 34 is **about** 5 × 30, or **150**.

2 × 289 is **about** 2 × 300, or **600**.

Find the best estimate.

1. 3 × 67
2. 2 × 889
3. 6 × 433
4. 8 × 356
5. 6 × 65
6. 3 × 931
7. 5 × 89
8. 9 × 22
9. 8 × 62
10. 3 × 75
11. 7 × 226
12. 8 × 580

Answer Box

A	B	C	D	E	F
3,200	210	1,400	420	2,700	1,800
G	**H**	**I**	**J**	**K**	**L**
240	450	2,400	480	180	4,800

Objective: Find the estimated product of a 2- to 3-digit number multiplied by a 1-digit number.

Break It into Pieces

Example

Find the product.

$$\begin{array}{r} 56 \\ \times\ 3 \\ \hline 18 \\ \underline{150} \\ \mathbf{168} \end{array} \quad \begin{array}{l} \\ \\ \leftarrow 3 \times 6 \\ \leftarrow 3 \times 50 \\ \\ \end{array}$$

$$3 \times 56 = (3 \times 6) + (3 \times 50)$$
$$= 18 + 150$$
$$= \mathbf{168}$$

Find the product.

1. $\begin{array}{r} 45 \\ \underline{\times\ 4} \end{array}$
2. $\begin{array}{r} 89 \\ \underline{\times\ 2} \end{array}$
3. $\begin{array}{r} 73 \\ \underline{\times\ 2} \end{array}$
4. $\begin{array}{r} 65 \\ \underline{\times\ 3} \end{array}$
5. $\begin{array}{r} 68 \\ \underline{\times\ 2} \end{array}$
6. $\begin{array}{r} 54 \\ \underline{\times\ 6} \end{array}$
7. $\begin{array}{r} 36 \\ \underline{\times\ 4} \end{array}$
8. $\begin{array}{r} 83 \\ \underline{\times\ 5} \end{array}$
9. $\begin{array}{r} 37 \\ \underline{\times\ 3} \end{array}$
10. $\begin{array}{r} 38 \\ \underline{\times\ 9} \end{array}$
11. $\begin{array}{r} 75 \\ \underline{\times\ 2} \end{array}$
12. $\begin{array}{r} 47 \\ \underline{\times\ 9} \end{array}$

Answer Box

A	B	C	D	E	F
146	111	324	178	195	180
G	**H**	**I**	**J**	**K**	**L**
136	150	415	423	144	342

Objective: Find the product of a 2-digit number multiplied by a 1-digit number.

Pick a Product

Find the product.

1. $\begin{array}{r} 56 \\ \underline{\times\ 4} \end{array}$
2. $\begin{array}{r} 94 \\ \underline{\times\ 3} \end{array}$
3. $\begin{array}{r} 38 \\ \underline{\times\ 2} \end{array}$
4. $\begin{array}{r} 53 \\ \underline{\times\ 4} \end{array}$
5. $\begin{array}{r} 66 \\ \underline{\times\ 3} \end{array}$
6. $\begin{array}{r} 41 \\ \underline{\times\ 7} \end{array}$
7. $\begin{array}{r} 32 \\ \underline{\times\ 4} \end{array}$
8. $\begin{array}{r} 22 \\ \underline{\times\ 6} \end{array}$
9. $\begin{array}{r} 38 \\ \underline{\times\ 6} \end{array}$
10. $\begin{array}{r} 73 \\ \underline{\times\ 2} \end{array}$

11. There are 45 noisemakers in each box.
How many are in 2 boxes?

12. Party hats are packaged 25 in each box.
You buy 3 boxes. How many party hats do you have?

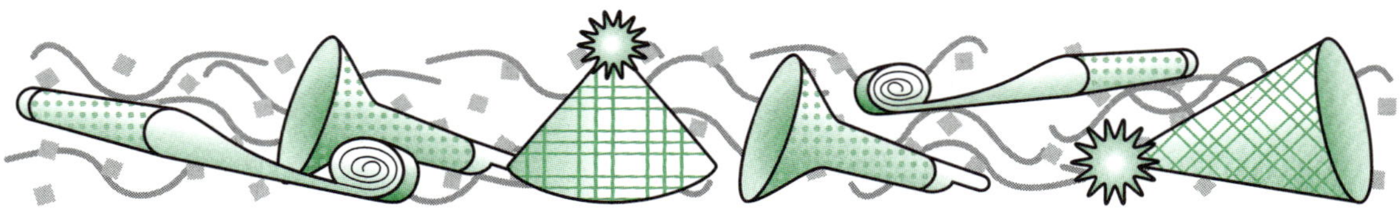

Answer Box

A	B	C	D	E	F
132	75	224	212	90	198
G	**H**	**I**	**J**	**K**	**L**
282	287	128	146	228	76

 Objective: Find the product of a 2-digit number multiplied by a 1-digit number.

Keep the Place in Place Value

Example

Find the product.

$$
\begin{array}{rl}
156 & \\
\underline{\times \quad 3} & \\
18 & \leftarrow 3 \times 6 \\
150 & \leftarrow 3 \times 50 \\
\underline{300} & \leftarrow 3 \times 100 \\
\mathbf{468} &
\end{array}
$$

Find the product.

1. 245×4
2. 843×2
3. 737×2
4. 652×3
5. 638×2
6. 354×6
7. 364×4
8. 483×5
9. 147×3
10. 348×9
11. 765×2
12. 756×3

Answer Box

A	B	C	D	E	F
2,124	2,415	1,956	980	1,276	441
G	**H**	**I**	**J**	**K**	**L**
2,268	1,686	1,456	3,132	1,474	1,530

Objective: Find the product of a 3-digit number multiplied by a 1-digit number.

From Ones to Hundreds!

Find the product.

1. $\begin{array}{r} 335 \\ \times \quad 4 \\ \hline \end{array}$

2. $\begin{array}{r} 843 \\ \times \quad 7 \\ \hline \end{array}$

3. $\begin{array}{r} 735 \\ \times \quad 6 \\ \hline \end{array}$

4. $\begin{array}{r} 631 \\ \times \quad 8 \\ \hline \end{array}$

5. $\begin{array}{r} 656 \\ \times \quad 4 \\ \hline \end{array}$

6. $\begin{array}{r} 154 \\ \times \quad 3 \\ \hline \end{array}$

7. $\begin{array}{r} 722 \\ \times \quad 3 \\ \hline \end{array}$

8. $\begin{array}{r} 480 \\ \times \quad 2 \\ \hline \end{array}$

9. $\begin{array}{r} 182 \\ \times \quad 3 \\ \hline \end{array}$

10. $\begin{array}{r} 948 \\ \times \quad 8 \\ \hline \end{array}$

11. There are 665 street intersections in Centerville. The town decides to put 4 wheelchair access ramps at each intersection. How many ramps is that altogether?

12. 406 people in Centerville each donate 7 h a week to maintaining the parks. How many hours per week is that in all?

Answer Box

A	B	C	D	E	F
2,660	4,410	1,340	5,048	5,901	2,842
G	**H**	**I**	**J**	**K**	**L**
546	2,624	7,584	2,166	960	462

 Objective: Find the product of a 3-digit number multiplied by a 1-digit number.

Money Multiplication

Find the total amount.

1. $\$2.97 \times 3$
2. $\$72.25 \times 2$
3. $\$88.95 \times 3$
4. $\$17.55 \times 8$
5. $\$20.30 \times 5$
6. $\$11.50 \times 8$
7. $\$58.15 \times 6$
8. $\$4.76 \times 2$
9. $\$47.25 \times 3$
10. $\$15.70 \times 2$
11. $\$25.30 \times 4$
12. $\$1.65 \times 6$

Answer Box

A	B	C	D	E	F
$9.52	$9.90	$348.90	$31.40	$101.50	$101.20
G	**H**	**I**	**J**	**K**	**L**
$92.00	$144.50	$140.40	$8.91	$141.75	$266.85

Objective: Find the product of an amount of money multiplied by a 1-digit number.

Problem Solving: Solving Two-Step Problems

The Art Shop	
Item	**Price**
picture frame	$7.89
watercolor paints	$8.50
sketch pad	$2.50
3 brushes	$10.95
10 pencils	$1.75
construction paper (20 sheets)	$5.99
tube of oil paint	$1.40
eraser	$0.75
clay	$6.95
glue stick	$1.35
box of crayons	$2.00
cardboard mat	$1.75
box of chalk	$1.50
easel	$35.99

Use the information in the table to find the answer.

1. What is the cost of 3 boxes of crayons and 2 sketch pads?
2. Which is more expensive, 6 tubes of oil paint or 3 boxes of crayons?
3. How much more does it cost to buy 2 packages of construction paper than a sketch pad?

4. Shawnee buys 6 tubes of oil paint and 3 brushes. How much does she spend?
5. Sue buys 3 boxes of chalk. How much change does she receive if she pays with a ten-dollar bill?
6. José has $6 to spend. Which can he buy, 6 tubes of oil paint or 3 boxes of crayons?
7. How much more does a picture frame cost than 2 sketch pads?
8. Ken wants to buy an easel and 4 tubes of oil paint. How much money does he need?
9. What does the art store charge for 6 brushes and a set of watercolor paints?
10. Mr. Ruiz needs 3 boxes of chalk for his classroom. Is $5 enough to buy them?
11. Joe buys 3 erasers and a sketch pad. How much does he spend?
12. Can Sarah buy 6 cardboard mats if she has $10?

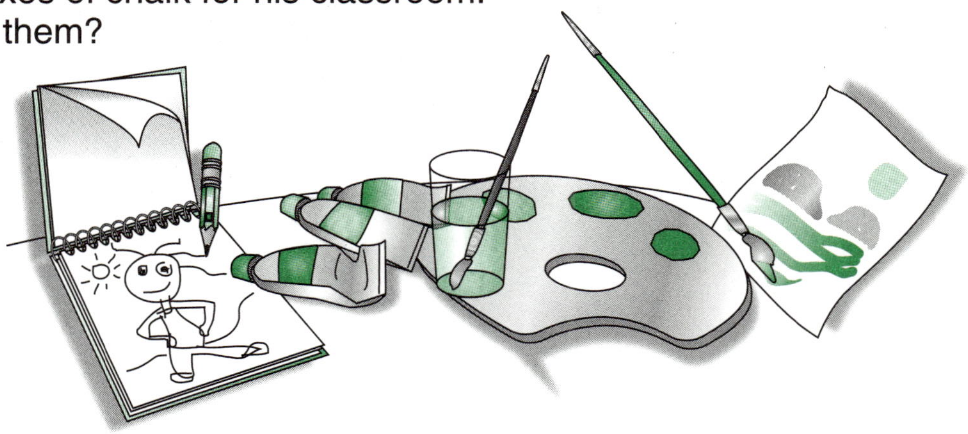

Answer Box

A	B	C	D	E	F
No	$41.59	$30.40	$9.48	Yes	$2.89
G	**H**	**I**	**J**	**K**	**L**
6 tubes of oil paint	3 boxes of crayons	$4.75	$5.50	$19.35	$11.00

Objective: Use a table to solve a two-step problem that involves addition, subtraction, or multiplication.

10 by 10

Example

Find the product.

$5 \times 2 = 10$, so $50 \times 20 = ■$

Think: **5** × **2** = **10**
50 × **2** = **100**
So, **50** × **20** = **1,000**.

Find the product.

1. $3 \times 4 = 12$, so $30 \times 40 = ■$
2. $3 \times 7 = 21$, so $30 \times 70 = ■$
3. $5 \times 6 = 30$, so $50 \times 60 = ■$
4. $7 \times 4 = 28$, so $70 \times 40 = ■$
5. $2 \times 9 = 18$, so $20 \times 90 = ■$
6. $5 \times 8 = 40$, so $50 \times 80 = ■$
7. $80 \times 40 = ■$
8. $60 \times 40 = ■$
9. $90 \times 30 = ■$
10. $40 \times 50 = ■$
11. A movie theater has 30 seats in each row. How many seats are there in 20 rows?
12. There are 40 rolls of theater tickets. If there are 20 tickets in each roll, how many tickets are there?

Answer Box

A	B	C	D	E	F
3,000	600	2,800	1,200	4,000	2,400
G	**H**	**I**	**J**	**K**	**L**
1,800	2,700	2,000	3,200	800	2,100

 Objective: Find a product of tens multiplied by tens.

100 by 10

Example

Find the product.

4 × 2 = 8, so 40 × 200 = ■

Think: **4** × **2** = **8**
40 × **2** = **8**0
40 × **2**0 = **8**00
So, **4**0 × **2**00 = **8**,000.

Find the product.

1. 2 × 4 = 8, so 20 × 400 = ■
2. 5 × 7 = 35, so 50 × 700 = ■
3. 3 × 7 = 21, so 30 × 700 = ■
4. 7 × 9 = 63, so 70 × 900 = ■
5. 2 × 8 = 16, so 20 × 800 = ■
6. 5 × 3 = 15, so 50 × 300 = ■
7. 30 × 600 = ■
8. 20 × 500 = ■
9. 90 × 400 = ■
10. 40 × 700 = ■
11. A carton holds 60 boxes of toothpaste. How many boxes are in 200 cartons?
12. There are 30 toothbrushes in each box. How many toothbrushes are in 200 boxes?

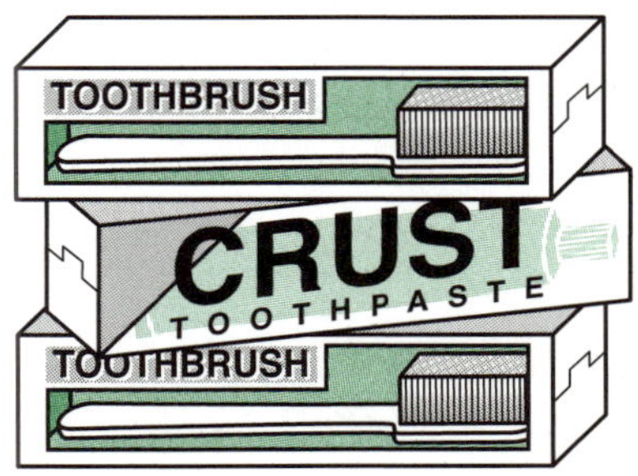

Answer Box

A	B	C	D	E	F
16,000	36,000	15,000	10,000	18,000	8,000
G	**H**	**I**	**J**	**K**	**L**
21,000	12,000	6,000	35,000	63,000	28,000

Objective: Find a product of hundreds multiplied by tens.

Estimating Tens

Example

Find the best estimate.

29 × 21 is **about** 30 × 20, or **600**.

34 × 57 is **about** 30 × 60, or **1,800**.

Find the best estimate.

1. 33 × 47
2. 12 × 88
3. 36 × 43
4. 87 × 35
5. 46 × 75
6. 35 × 73
7. 52 × 89
8. 39 × 62
9. 88 × 62
10. 73 × 75
11. 97 × 22
12. 68 × 57

Answer Box

A	B	C	D	E	F
1,500	3,600	4,200	900	5,400	1,600
G	**H**	**I**	**J**	**K**	**L**
4,500	5,600	2,800	2,400	4,000	2,000

 Objective: Find the estimated product of a 2-digit number multiplied by a 2-digit number.

Estimating Tens and Hundreds

Example

Find the best estimate.

19 × 621 is **about** 20 × 600, or **12,000**.

63 × 487 is **about** 60 × 500, or **30,000**.

Find the best estimate.

1. 77 × 147	2. 92 × 848	3. 56 × 543
4. 57 × 935	5. 66 × 785	6. 25 × 193
7. 42 × 689	8. 79 × 632	9. 38 × 362
10. 72 × 745	11. 27 × 822	12. 98 × 857

Answer Box

A	B	C	D	E	F
28,000	8,000	16,000	30,000	6,000	48,000
G	**H**	**I**	**J**	**K**	**L**
54,000	49,000	56,000	24,000	90,000	72,000

Objective: Find the estimated product of a 3-digit number multiplied by a 2-digit number.

Two by Two

Example

Find the product.

$$\begin{array}{r} 73 \\ \times\ 18 \\ \hline 584 \\ +\ 730 \\ \hline \mathbf{1,314} \end{array} \quad \begin{array}{l} \\ \\ \leftarrow 8 \times 73 \\ \leftarrow 10 \times 73 \\ \\ \end{array}$$

Find the product.

1. 56×34
2. 94×33
3. 38×62
4. 62×94
5. 66×52
6. 41×57
7. 32×14
8. 94×45
9. 38×76
10. 73×22
11. 45×32
12. 37×55

Answer Box

A	B	C	D	E	F
2,888	2,356	5,828	1,904	2,337	3,102
G	**H**	**I**	**J**	**K**	**L**
3,432	1,440	448	1,606	4,230	2,035

Objective: Find the product of a 2-digit number multiplied by a 2-digit number.

More Two by Two

Find the product.

1. 69 × 74
2. 46 × 85
3. 32 × 76
4. 84 × 28
5. 67 × 37
6. 93 × 27
7. 88 × 12
8. 47 × 86
9. 43 × 52
10. 62 × 28
11. The bus that tours the Colonial Village Park holds 45 passengers. If the bus makes 8 runs per day, then how many passengers can it carry in 8 days?
12. The bus tours Colonial Village Park 5 times a day. In one year, the bus ran only for 36 weeks. How many times did the bus tour in that year? (Hint: There are seven days in a week.)

Answer Box

A	B	C	D	E	F
1,260	4,042	5,106	2,352	2,479	2,880
G	**H**	**I**	**J**	**K**	**L**
2,511	3,910	1,736	1,056	2,432	2,236

Objective: Find the product of a 2-digit number multiplied by a 2-digit number.

Two by Three

Find the product.

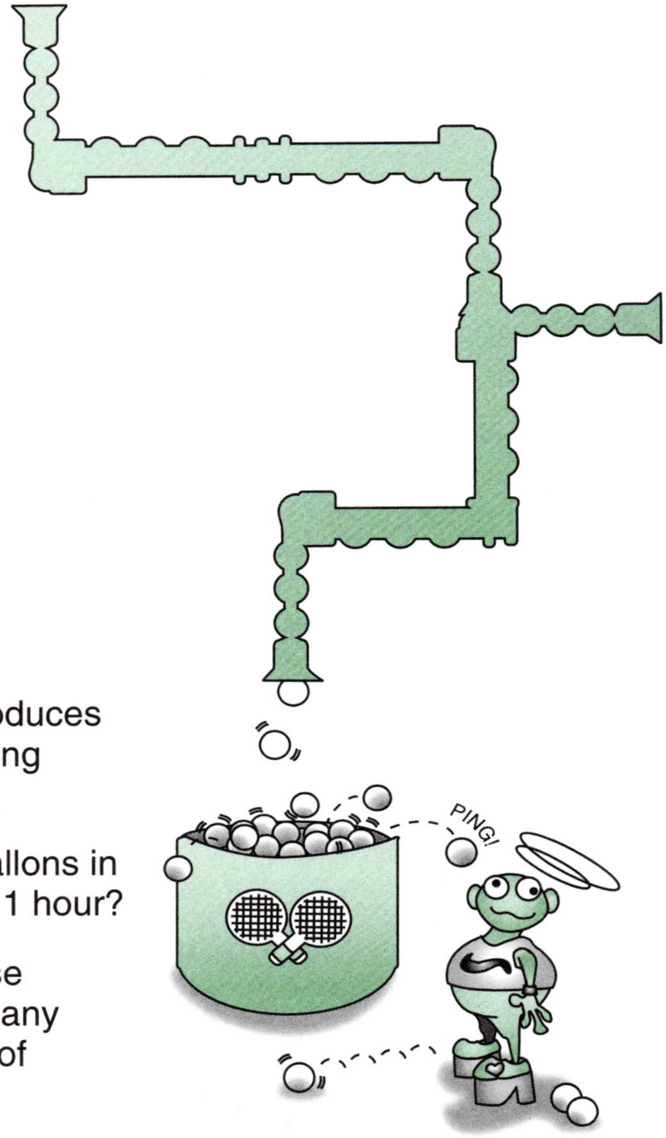

1. 405×68
2. 810×42
3. 833×73
4. 376×94
5. 274×19
6. 887×56
7. 634×68
8. 527×37
9. 642×43
10. Each hour the Table Tennis Factory produces 361 ping pong balls. How many ping pong balls are produced in 3 days?
11. A milk producer's machinery fills 993 gallons in 1 minute. How many gallons can it fill in 1 hour?
12. One machine packages 126 lb of cheese each day. The Farmland Cheese Company owns 87 machines. How many pounds of cheese can they package each day?

Answer Box

A	B	C	D	E	F
25,992	35,344	27,606	60,809	34,020	10,962
G	**H**	**I**	**J**	**K**	**L**
27,540	43,112	5,206	59,580	19,499	49,672

 Objective: Find the product of a 3-digit number multiplied by a 2-digit number.

More Two by Three

Find the product.

1. 579×45	2. 622×34	3. 106×43	4. 950×68	5. 264×96
6. 771×63	7. 601×77	8. 400×91	9. 388×11	10. 477×21

11. About 680 shoppers eat lunch in the food court at the mall each day. How many people eat lunch in the food court in 7 weeks? (Hint: There are seven days in a week.)

12. An ice cream stand sells about 430 ice cream treats each day during the summer. The ice cream stand is open for 12 weeks each summer. How many ice cream treats are sold?

Answer Box

A	B	C	D	E	F
25,344	4,268	10,017	46,277	48,573	36,400
G	**H**	**I**	**J**	**K**	**L**
4,558	33,320	64,600	26,055	36,120	21,148

Objective: Find the product of a 3-digit number multiplied by a 2-digit number.

Problem Solving: Choosing the Operation

Choose the number sentence to solve the problem.

1. The fourth grade class went on a trip to a nature center. Two buses each carried 37 students. How many students went on the trip?

   2 × 37 = ■ or 2 + 37 = ■

2. Outside the center, Kelly counted 12 birds. Devon saw 16 birds. How many more birds did Devon see than Kelly?

   12 + 16 = ■ or 16 – 12 = ■

3. At the gift shop, postcards are $0.75 each. Sam buys 5 cards. How much does he spend?

   5 – $0.75 = ■ or 5 × $0.75 = ■

4. The teacher buys a book that costs $13.50. He gives the cashier $20. How much change does he receive?

   $20 – $13.50 = ■ or $20 + $13.50 = ■

5. Tomás and David each bring a bag of carrots for a snack. Tomás has 12 carrots in his bag. David has twice as many. How many pieces does David have?

   2 × 12 = ■ or 12 + 2 = ■

6. Jamie saw 7 butterflies in the butterfly garden, while Leon saw 9. How many more did Leon see than Jamie?

   9 × 7 = ■ or 9 – 7 = ■

7 The class went on a nature walk after lunch. One group walked 0.3 miles and the other group walked 0.7 miles. If both groups left at 2 o'clock, then what time did the first group return?

0.3 + 0.7 = ■ or not enough information

8 Alicia went on the frog hunt and saw 2 bullfrogs and 3 leopard frogs. Nate found 3 green frogs. How many frogs did they see in all?

2 + 3 + 3 = ■ or 2 × 3 = ■

9 John collected 24 different leaves for his collection. If he had an equal number at home already, how many leaves did he have after the trip?

2 × 24 = ■ or not enough information

10 Helen took 3 rolls of film on the trip. If each roll had 12 pictures on it, then how many pictures did she take?

3 × 12 = ■ or 3 + 12 = ■

11 In one set of 24 pictures, there were 18 pictures of birds and frogs. The rest of the pictures showed scenery. How many pictures of scenery were on the roll?

18 + 24 = ■ or 24 – 18 = ■

12 Doug brought a canteen with 2 cups of water in it. His 3 friends each brought the same amount of water on the trip. How much did they have to drink altogether?

2 + 3 = ■ or 2 × 4 = ■

Answer Box

A	B	C	D	E	F
2 × 37 = ■	Not enough information	2 + 3 + 3 = ■	2 × 4 = ■	2 × 24 = ■	2 × 12 = ■
G	**H**	**I**	**J**	**K**	**L**
3 × 12 = ■	\$20 – \$13.50 = ■	16 – 12 = ■	9 – 7 = ■	24 – 18 = ■	5 × \$0.75 = ■

Objective: Find the addition, subtraction, or multiplication number sentence to solve a problem.

How Much Money Is That?

Find the product.

1	$135 × 74	2	$846 × 55
3	$335 × 96	4	$601 × 58
5	$456 × 64	6	$159 × 43
7	$122 × 63	8	$635 × 84
9	$582 × 33	10	$706 × 47

11 A computer store sells computers for $948 each. The store has 68 computers for sale. If all the computers are sold, what will be the total sales?

12 A computer show sells full-day passes for $22 each. On one day, they sold 400 full-day passes. How much money did they make?

Answer Box

A	B	C	D	E	F
$32,160	$19,206	$53,340	$8,800	$9,990	$34,858
G	**H**	**I**	**J**	**K**	**L**
$64,464	$29,184	$6,837	$46,530	$7,686	$33,182

 Objective: Find the product of an amount of money multiplied by a 2-digit number.